WITHDRAWN

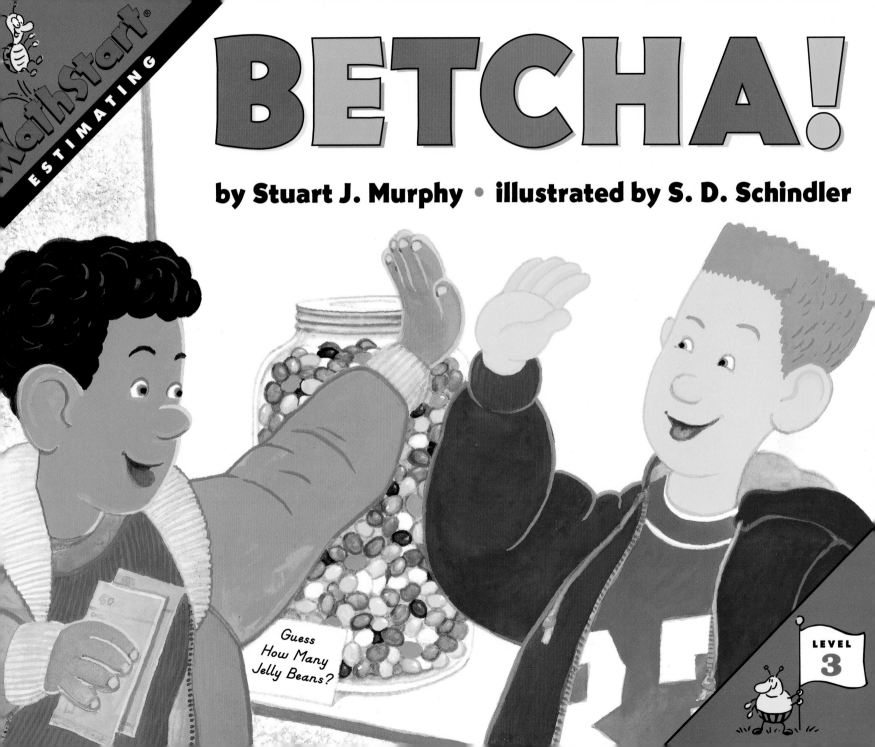

Be sure to look for all of these **MathStart** books:

US $4.95 / $6.95 CAN
ISBN 0-06-446707-4

9 780006 467070

50495

Cover art © 1997 by S. D. Schlindler
Author photo by Stuart Rodgers Photography, Ltd.

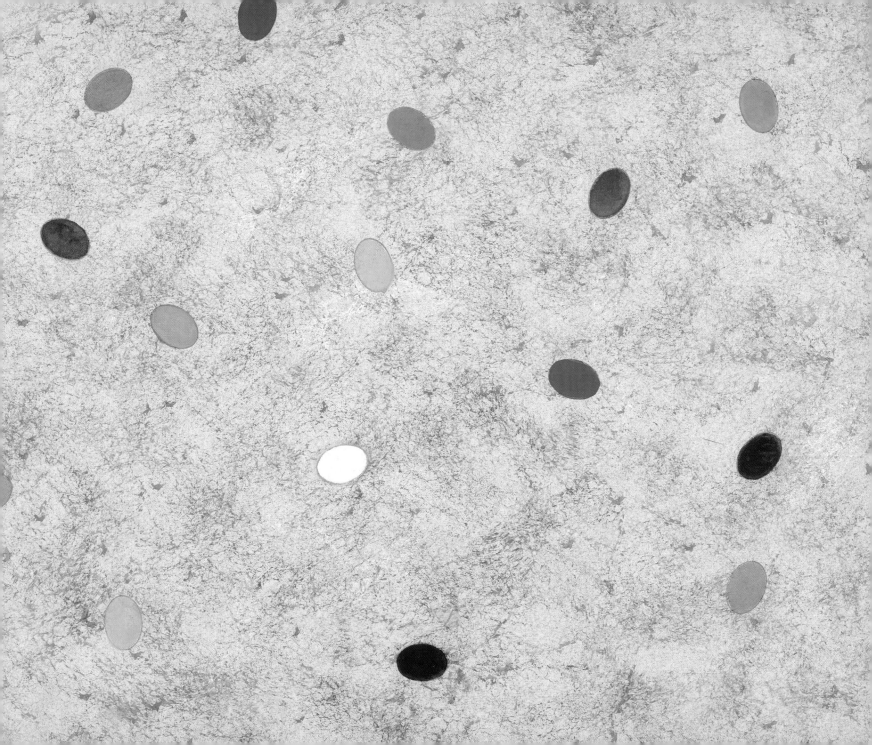

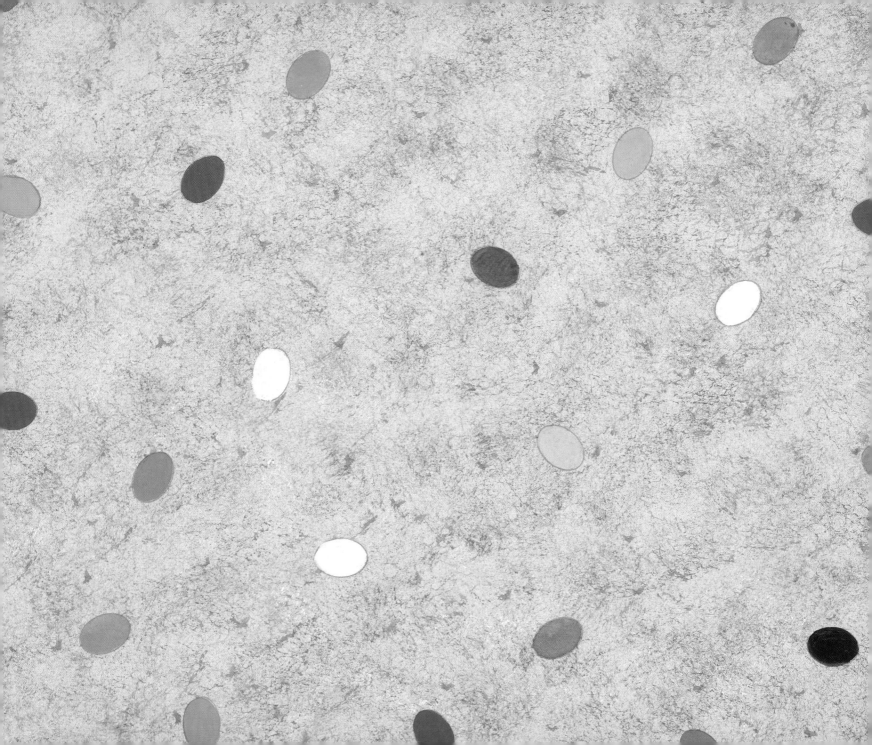

MathStart™
ESTIMATING

BETCHA!

by Stuart J. Murphy • illustrated by S. D. Schindler

HarperCollins*Publishers*

LEVEL
3

**To Katherine Brown Tegen—
who was willing to betcha
that MathStart would work.**

—S.J.M.

For more information about the MathStart series, please write to
HarperCollins Children's Books, 10 East 53rd Street, New York, NY 10022,
or visit our web site at http://www.harperchildrens.com.
Bugs incorporated in the MathStart series design were painted by Jon Buller.
HarperCollins®, 🐛®, and MathStart™ are trademarks of HarperCollins Publishers Inc.

BETCHA!

Library of Congress Cataloging-in-Publication Data
Murphy, Stuart J., date.
    Betcha! / by Stuart J. Murphy ; illustrated by S. D. Schindler.
        p.   cm. — (MathStart)
    "Level 3, Estimating."
    Summary: Uses a dialog between friends, one who estimates, one who counts precisely, to show
estimation at work in everyday life.
    ISBN 0-06-026768-2. — ISBN 0-06-026769-0 (lib. bdg.). — ISBN 0-06-446707-4 (pbk.)
    1. Estimation theory—Juvenile literature. [1. Estimation theory. 2. Arithmetic.] I. Schindler,
S. D., ill. II. Title. III. Series.
QA276.8.M87   1997   96-15486
519.5′44—dc20   CIP
                                                                                      AC

Typography by Elynn Cohen
4   5   6   7   8   9   10
❖

BETCHA!

Hey! Look at this. Planet Toys is having a contest. Whoever guesses the correct number of jelly beans in the jar in their window wins two free tickets to the All-Star Game!

PLANET TOYS

Guess how many jelly beans are in me!

Oh, yeah?
Let's go down and check it out.

Betcha I can win.

Oh yeah? I'm really good at figuring things out.
I betcha that I'll win.

If you're so good at figuring things out, can you tell me how many other people are on this bus?

You betcha!

Okay. How many?

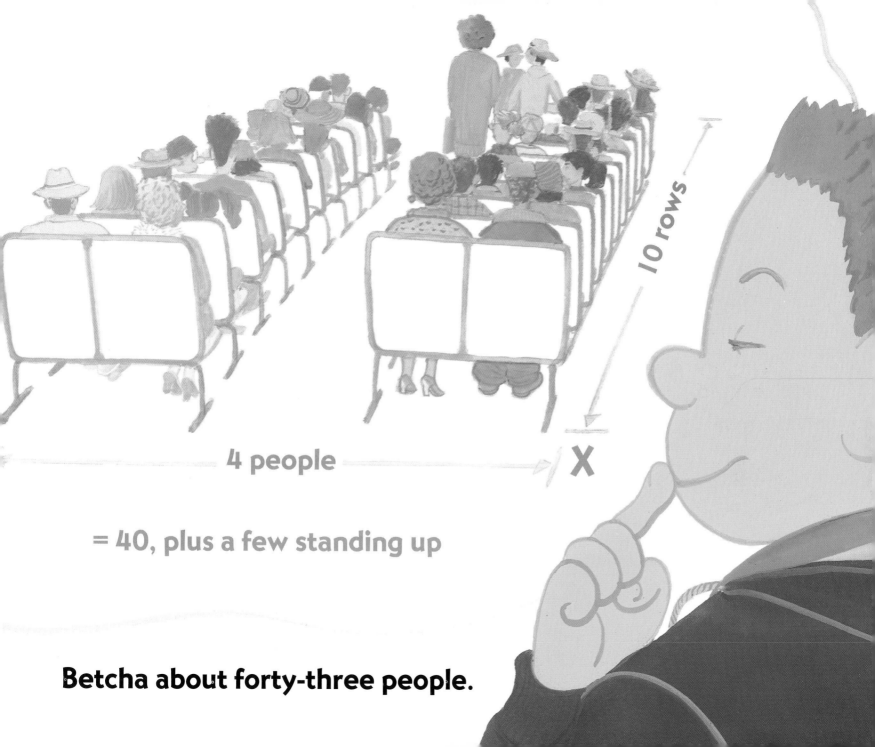

4 people ⟶ X
10 rows

= 40, plus a few standing up

Betcha about forty-three people.

I just counted, and I got forty-five.

I was pretty close!

Wow! Look at that traffic jam.
Betcha you can't tell me how many cars are
stuck on the block.

Betcha I can.

Go ahead and try.

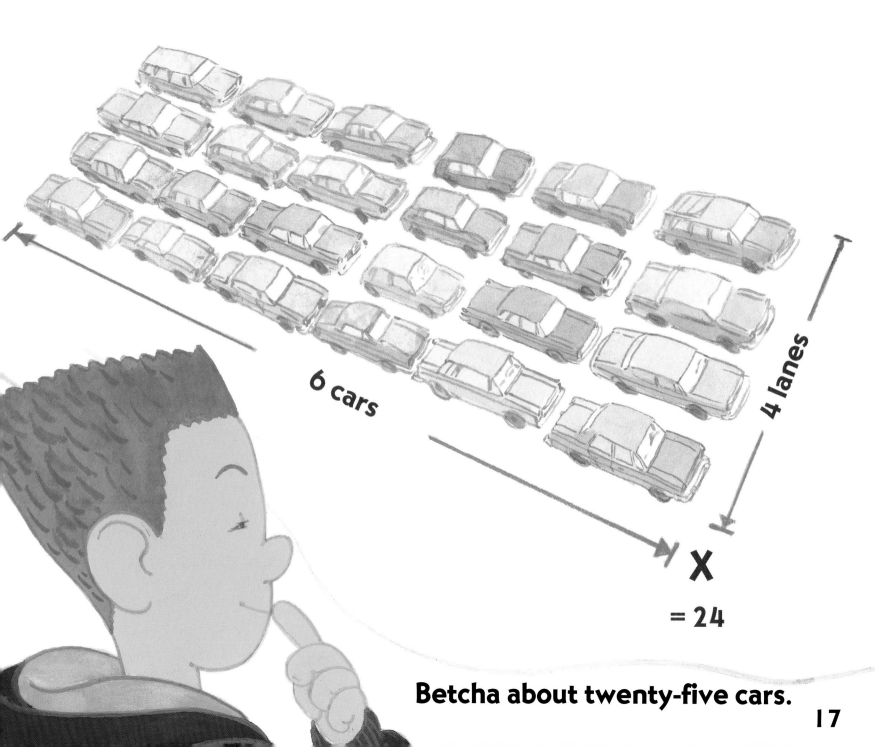

6 cars

4 lanes

X

= 24

Betcha about twenty-five cars.

17

I counted twenty-three.

Here's the store. Look at all that cool stuff! About how much do you think it would cost to buy it all?

PLANET TOYS

$39

WALKIE-
TALKIES

$28

Turbo-Fire

$22

$12

Betcha I can tell you about how much.

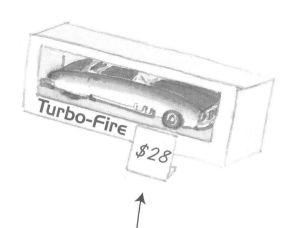

$39

$22

$28

Almost $40 + just over $20 = about $60

$60 + almost $30 = about $90

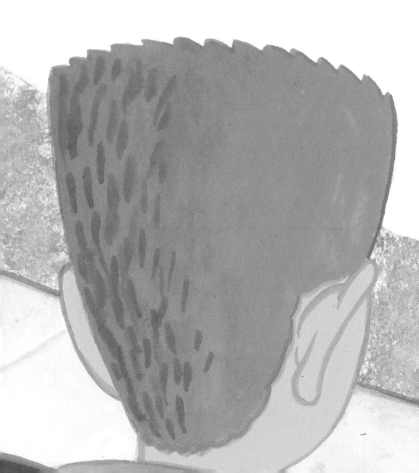

**$90 + just over $10
= about $100**

$12

Betcha about $100.

23

See? I was close again,
and I didn't even need a pencil.

Okay. Now for the real thing.
About how many jelly beans are in the jar?

Guess how many jelly beans are in me and win two free tickets to the All-Star Game?

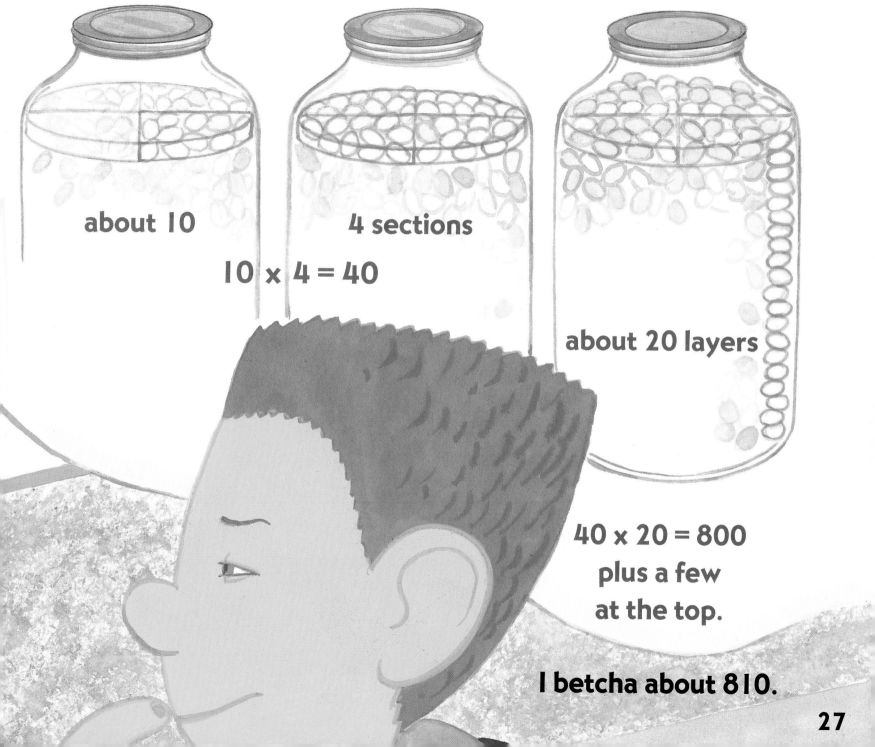

about 10

4 sections

$10 \times 4 = 40$

about 20 layers

$40 \times 20 = 800$
plus a few
at the top.

I betcha about 810.

27

Incredible! That's exactly right!
You're headed to the All-Star Game.

Betcha I know who wants the other ticket!

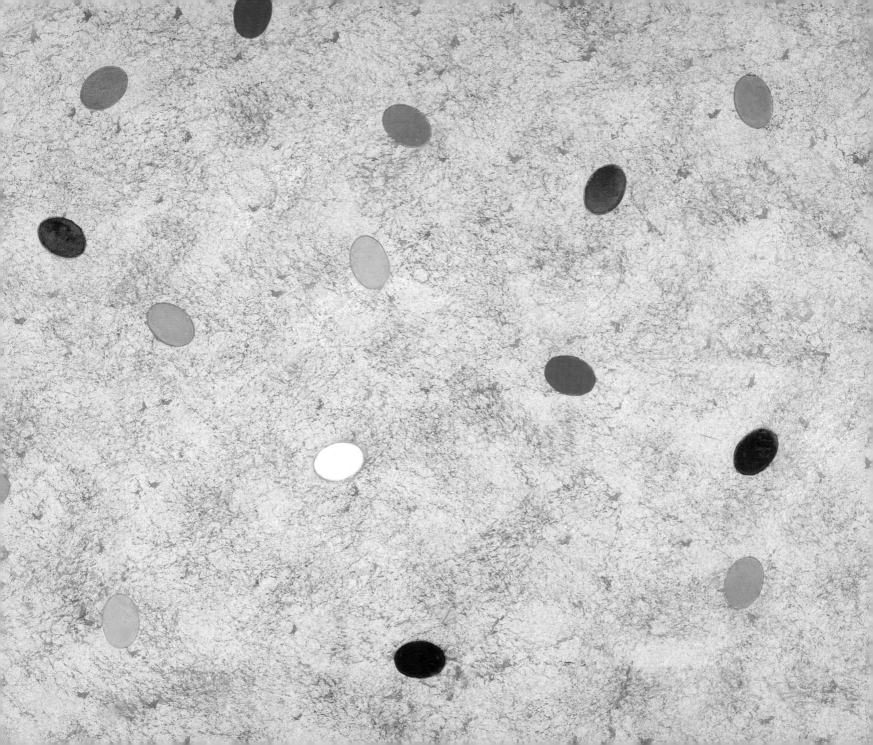

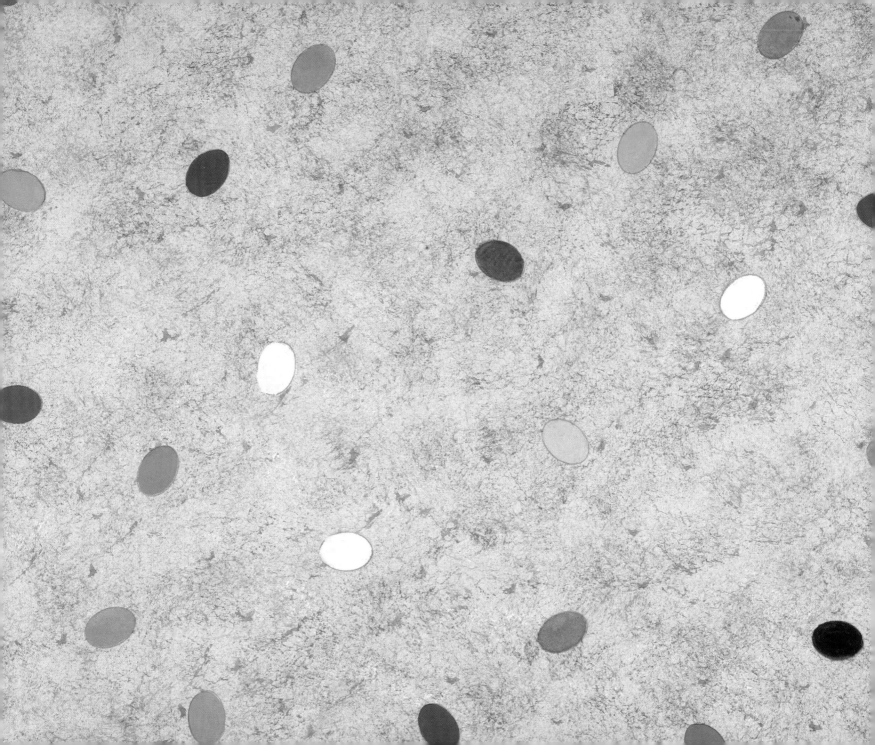

6405

STUART J. MURPHY is a visual learning specialist. A graduate of the Rhode Island School of Design, he has a strong background in design and art direction. He also has extensive experience in the world of educational publishing. Drawing on all these talents, Stuart Murphy brings a unique perspective to the **MathStart** series. In **MathStart** books pictures do more than tell stories; they teach math.

Stuart Murphy lives in Evanston, Illinois. He always plays "Betcha how much the check is!" in restaurants.

S. D. SCHINDLER has illustrated many popular books for children, including SPOOKY TRICKS by Rose Wyler and Gerald Ames, CATWINGS by Ursula LeGuin, BIG PUMPKIN by Erica Silverman, and IS THIS A HOUSE FOR HERMIT CRAB? by Megan McDonald, which won the International Reading Association's Children's Book Award.

S. D. Schindler lives in Philadelphia, Pennsylvania.

MathStart®

STUART J. MURPHY travels all over the United States talking to thousands of kids. And you'll never believe what they talk about: MATH! Stuart shows kids that they use math every day—to share a pizza, spend their allowance, and even sort socks. Stuart writes funny stories about math—and if you read his books, you'll start to see the fun in math, too.

BETCHA!

Betcha you'll learn to estimate as two friends guess how many people are riding their bus, how much a bunch of toys costs, and how many jelly beans fill a jar. Betcha you'll have fun, too!

LEVEL 1 — Ages 3 up
LEVEL 2 — Ages 6 up
LEVEL 3 — Ages 7 up

Visit MathStart at http://www.harperchildrens.com.

Ages 7 up
HarperTrophy®

US $4.95 / $6.95 CAN
ISBN 0-06-446707-4

46707

0 46594 00495 6

BETCHA!
by Stuart J. Murphy · illustrated by S. D. Schindler

TOO MANY KANGAROO THINGS TO DO!
by Stuart J. Murphy
illustrated by Kevin O'Malley

DIVIDE and RIDE
by Stuart J. Murphy
illustrated by George Ulrich